PRINCIPLES OF ALGEBRA
A
Fundamental Approach

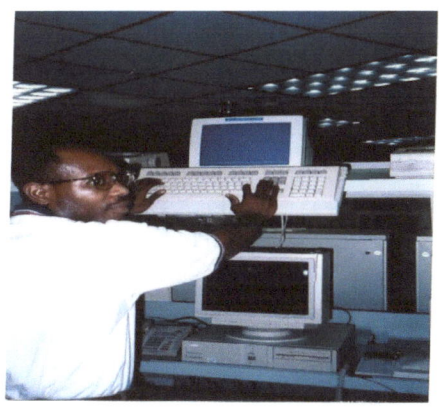

CREATING ALGORITHMS AND FUNCTIONS

By Dr. Iwasan D. Kejawa, Ed. D

INTRODUCTION

To create system Algorithms one must represent the objects and the functions with variables. Algorithms are used to analyze various businesses, science and engineering problems as well as other problems encounter in our daily lives. We often create formula by our actions which we are not aware off. Our activities or actions can be represented systematically by Algorithms. We do apply Algorithms in everything that we do. The systems by which we do things are Algorithms: Formula can be categorized as written plan to solve a problem or to accomplish a function. It is usually in Mathematical or Algorithmic form.

FOR INSTANCE:

If 16 items are to be sold for 5 dollars each and 3 of the items are sold for 7 dollars each. What is the total price of all the items? How many items remain if only 10 of the items are sold?

Representations:

Let A = the number of item1

B = the number of item2, etc.

Z = Remainder

S = number of items sold

Then Algorithms:

$T = AP$ or $A1 \times P1$

Where A = number of each item1

And B = Price of each item1

$Z = A - S$

A = OBJECTS, P and S are functions. And

$P(A) = T, P(S) = Z$

T and Z are outcomes or end results.

Solving for T and Z:

$$A = 16$$

P1 is the first instance which is equal 5

$$P(1) = 5, P(2) = 7$$

P2 is the second instance which is 7

$$S = P(1) + P(2) = 12$$

$$S = 12$$

T1 is the total of first instance and

T2 is the total of second instance

$$T1 = (16 - 3) \times 5 = \$65$$

$$T2 = 3 \times 7 = \$21$$

$$T = 65 + 21 = \$86$$

$$Z = A - S$$

$$= 16 - 10$$

$$= 6$$

MATHEMATICAL FUNDAMENTALS

Order of Performance in Mathematics

To perform calculations in mathematics, it must be done by order of performance or operation:

For example:

$A + B(C-D)4 - Z$

$(C - D)$ ---------- First

$(C-D)4$ --------------- Second

$B(C-D)4$ --------------- Third

$A+B(C-D)4$ ---------------- Fourth

$A+B(C-D)4+Z$ --------------- Fifth

The acronyms for this example are the followings:

PEMDSA (Please Excuse My Dear Sister Anne)

PEMDAS (Please Excuse My Dear Aunt Sally)

        Or

PEMDSA (Please Expect Me December Sunday Afternoon)

MATHEMATICAL FUNDAMENTALS 8

PEMDAS (Please Expect Me December At Sun done)

PERFORMANCE EVALUATION ALGORITHMS

To create and calculate performance we will multiply activity by action.

Performance evaluation can also be regarded as combinations of action and object

Formula Representation:

Let p = Performance, Action = A and Activity = B

Then Formula:

$$P = A + B$$

For instance:

Imagine an automobile that travels at a constant speed of 80 miles per hour and it took 5 hours for the driver to get to his or her designated destination. What is the overall performance of the automobile.

Since A= action and B activity

Activity = 80

Action = 5

Then

Overall Performance = 80 x 5

= 400 miles per 5 hours

Or to put another way,

We say:

Rate x Time = Distance

80mph x 5hr = 400

MATHEMATICAL FUNDAMENTALS

FUNCTIONS

ADDITION FUNCTION

Addition is the combination of two or more entities.

Formula Representation:

$$A + B + \ldots\ldots n$$

Where n varies

For instance

Let assume we want to add six numbers OR more.

Then Formula:

$$\text{Function (F1)} = n1 + n2 + n3 + n4 + n5 + n6 + \ldots N$$

$$(F2) = (n1, n2, n3, n4, n5, n6, n)$$

Example:

Subtraction Function

Subtraction is the elimination of one or more variables from an entity

MATHEMATICAL FUNDAMENTALS

FORMULA REPRESENTATION:

$$A - b - \ldots N$$

FOR INSTANCE

We can eliminate or take away several entities or items from a Whole.

Thus Formula:

$$\text{Function } (F) = N - n_1 - n_2 - n_3 - \ldots n$$

Example: To subtract the number 23 and 25 from whole number 100.

$$\text{Function } (F) = 100 - 23 - 25$$
$$= 52$$

MULTIPLICATION FUNCTION

Multiplication is the addition of a variable *n* number of times.

For instance

The addition of Y *n* number of times

MATHEMATICAL FUNDAMENTALS

$$Y(n)$$

FORMULA REPRESENTATION:

    Then Algorithms:

        Function (F) = Y x n

        Y * n

        Y. n

= Y1 + Y2 + Y3 + ... + Yen

Example: To add a number 1000, 1250, 515 and 6000 series of time

    Function (F) = 1000 + 1250 + 513 + 6000 = 8763

DIVISION FUNCTION

The division function is the partition of a whole (Y) of entity n into x number of times. It is the number of times n occurs in Y or the occurrence of n at x number of times.

FORMULA REPRESENTATION:

    FUNCTION (F1) = Y / n

Thus Formula:

    (F1) = n (1), n (2), n (3), n(x)

        (F1) = x

Example:

To divide the number 12 by 2 will be

$F = 12 / 2$

$F = 2(1), 2(2), 2(3), 2(4), 2(5), 2(6)$

$F = 6$

QUOTIENT FUNCTION

Quotient function is how many times an entity varies after the entity is divided by its divisor,

For Instance

Let assume N is divided by y at z times

FORMULA REPRESENTATION:

Then Algorithms:

$\text{FUNCTION}(F) = N / y(1) . N / y(z)$

Where $Y(z)$ not equal 0

Thus:

$$y = \text{divisor of N such that } y_1 \times y_2 \ldots y_z = N$$

$$z = \text{occurrence of } y$$

FUNCTION (F) = z

Example:

To find the quotient of a number 8

The divisors 1, 2, 4 and 8

Since $8/4 = 2$ and $2 \times 2 \times 2 = 8$

Using the above that was given earlier, then:

Function (F) = $8/4_{(1)} \times 8/4_{(2)} \times 8/4_{(3)}$

Quotient = 3

AVERAGE OR MEAN FUNCTION

An average function is the sum of a group of numbers divided by the number of times the number existed in the sum. The average function of six numbers represented as A, B, C, and D, E, F would be for example being formulated as follows:

FORMULA REPRESENTATION:

Formula:

Average Function (F) = $(A_1 + B_2 + C_3 + D_4 + E_5 + F_N) / n$

A, B, C, D, E, F = set of numbers

n = number of occurrence

Example: A

An average function of a group of numbers 2, 4, 1, 3, 2

First count the number occurrence of the number group:

$y(1) = 2, y(2) = 4, y(3) = 1, y(4) = 3, y(5) = 2$

n = 5

Thus average Function

$(F) = (2(1) + 4(2) + 1(3) + 3(4) + 2(5)) / 5 = 12/5$

$(F) = (2 + 4 + 1 + 3 + 2) / 5$

$= 12 / 5$

Average Function (F) = 2.4

MATHEMATICAL FUNDAMENTALS

ROOT FUNCTION

Root function is number of times a divisor of giving number is multiply by itself to equal to the whole number.

Representation Formula:

$F = N/n(y) \times N/n(y)$

N = the giving number

n = the divisor,

y = number of times it occur

Example A:

To find the root function of the number 16

First find all the possible divisors, then

How many times it is multiply by itself to arrive at the same number?

Function (F) = $16/4(1) \times 16/4(2)$

$= 4 \times 4$

$= 16$

$y = 2,$

$n = 4$,

Therefore we say that 4 is the square root 16

Since when it is multiply by itself yield 16

Alternatively:

Function (F=) $16/8(1) \times 16/8(2) \times 16/8(3) \times 16/8(4)$

$= 2 \times 2 \times 2 \times 2$

$= 16$

$y = 4$

$n = 2$

We arrive at the same solution except that y and n are Inter-changed. We may say that 2 is the fourth root of 16

Example B:

If we are ask to find root function of 27?

$N = 9$, the divisor = y,

Number of occurrence = n

Then the formula =

$N/y (1).N/y (2).N/y (3)$ and so on .... (n)

Therefore,

= 9/3(1). 9/3(2). 9/3(3)

   3 . 3 . 3

= 27

y = 3

n = 3

Therefore the root is equal to: 3

Here we say 3 is the cube root of 27

Example C: Find the root number of 36?

Solution: N = 36, possible divisors is 9, 3, 4, 6

Therefore y = 9, 3, 4, 6

When y = 9

Then using the formula, it is 36/9(1). 36/9(2). 36/9(3)

Which equals?

y = 9

= 4. 4. 4

= 64

Therefore 4 is not the root of 36 because 64 is large that the actual number. (36)

When y = 3

We have 36/3(1). 36/3(2) 36/3(3)

= 12. 12. 12

= 1728

Since 1728 is larger than 36

Then 9 is not the root of 36 because 729 is larger than 3

When y = 6

We have:

= 36/6(1). 36/6(2). 36/6(3)

= 3. 3 . 3

= 27

y = 3

n = 6

Since 27 is less than 36 then 4 is not the root

of 36

When y = 4

We have:

= 36/4(1). 36/4(2)

= 6. 6

= 36

Since this is equal to actual number then 6 is the

root of 36, we may say that 6 is the square root of 36 because when 6 is multiply itself we get the whole number

36

MATHEMATICAL FUNDAMENTALS

CHAPTER TWO

EQUATIONS

Equations can be categorized as (1) Simple Equation, (2) Simultaneous Equation (3) Quadratic Equation, (4) Monomial Equation (5) Binomial Equation and (6) Polynomial Equation. These equations can be regarded also as functions:

- Simple Function
- Simultaneous Function
- Quadratic Function
- Monomial Function
- Binomial Function
- Trinomial Function
- Polynomial Function

Simple Equations

These types of equations can be created by multiplying, dividing, subtracting or adding fixed numbers and variables. A simple equation function may be in the following terms, a fixed number n and variables y an x:

Representations:

i. $F = ny$

ii. $F = n + y$

iii. $F = n/(n + y)$

iv. $F = n - y$

v. $F = n - y - x$

The above examples are just possibilities of simple equation functions. There are a lot of simple equation functions.

Example: A

Let a given variable number be y and a fixed number be 8 and a function be equal 16. Now we can derive the appropriate possible simple equations.

$n = 8$

y=variable number

$F = 16$

The equations for the above example may be as follows:

1) $16 = 8y$

   or

2) $8y = 16$

3) $8y/8 = 16/8$

4) $y = 2$

5) $16 = 8 + y$ or $y + 8 = 16$

6) $y = 16 - 8$

7) $y = 8$

8) $16 = 8/(8+y)$

MATHEMATICAL FUNDAMENTALS

$$16(8 + y) = 8$$

$$128 + y = 8$$

$$Y = -120$$

9) $16 = 8 + y$

10) $-y = 16 - 8$

11) $-y = 8$

12) $y = -8$

With example A, let another fixed number (x) to be 3, then

13) $16 = 8 - y - 3$

14) $16 - 8 + 3 = -y$

15) $8 + 3 = -y$

16) $11 = -y$

17) $y = -11$

Example B:

Let a given variable be equal to 45 and a fixed number equal 65 with given function 3. What are possible simple equations?

$T = 65$

$r = 45$

$z = 3$

The possible number of equations for the above example can be enumerated as follows:

1). $45r + 3z = 65$

2). $3z = 65 - 45r$

3). $T = \dfrac{65 - 45r}{z}$

Simultaneous Equations

These types of equations are formed with two different or same types of functions. Each function can be in form of two equations. Both of the equations may also be in form of additions, subtractions, multiplications and divisions.

Representations:

i. $x + y = C$
    $x - y = Z$

ii. $x - y = C$

$$x - y = Z$$

iii. $$Ax - Ey = C$$
$$Ax + By = Z$$

iv. $$Ax - Ey = C$$
$$Ax + Ey = Z$$

v. $$Ax/By = C$$
$$Ax/By = z$$

vi. $$Ax = C$$
$$Ey = Z$$

Example A:

If in a process, Fl = 10 and if a given number 4 is multiply by a variable x and added to a fixed number 2 which is multiply by a variable y. In another process F2 = 8 and a given fixed number 2 is multiply by a variable x and subtracted from a given fixed number 5 which is multiply by a variable y. Formulate the appropriate simultaneous equations for this scenario and solve for x and y

Representations:

$F1 = 4x + 2y = 10$ i

$F2 = 2x + 5y = 8$ ii

Solving:

In other to arrive at a solution we can multiply equation ii by (- 2) as follows:

$4x + 2y = 10$.........i

$(-2(2x)) + (-2(5y)) = (-2(8))$ which will be equal to:

$4x + (-10y) = -6$ OR $-4x - 8y = -6$

Both equations can be represented as follows:

$4x + 2y = 10$..................i

$-4x - 10y = -6$..................ii

We can now eliminate x and solve for y by adding equation is from equation ii.

Quadratic Equations

MATHEMATICAL FUNDAMENTALS

These kinds of equations can be termed as functions where a variable is multiply by itself once or more number of times in the equation (i.e varies once or more times) and combined with a constant number or fixed number (Refer to the section on Root Functions on Chapter One).

Representations:

$F1 = Z_{(1)} \times Z_{(n\ times)} + N = 0$

$F2 = Z.Z_{(n)\ times} + N = 0$

$F3 = Z' + N$ where' equals n times

$F4 = Z * Z_{(n)\ times} + N = 0$

Z, n are variables While N is constant.

Example A

Let us look at a function that has a variable number of 6 which varies two times and added to a Constant number 3. What is the solution for the Function and its quadratic equations Algorithms?

Solving

Let $Z = 6$

$n = 2$

$N = 3$

Therefore

$F = Z_{(1)} \times Z_{(2)} + 3 = 0$

$= 6 \times 6 + 3 = 0$

$F = 15$

Or

$$F = Z.Z + 3 = 0$$
$$= 6.6 + 3 = 0$$
$$F = 15$$

Or

$$F = Z' + 3 = 0 \text{ where}' = 2$$
$$6' + 3 = 15$$

Or $\quad 6 * 6 + 3 = 15$

Polynomial Equations

The Binomial Equations are equations where a variable is multiply by self n number of times (that is (i.e.) n times) combined with a variable and constant or fixed number

Representations:

$$F1 = Z_{(n)} \times Z_{(n)} + T + N$$

$$F2 = Z_{(1)}.Z_{(n)} + T + N$$

$$F3 = Z' + T + N$$

$$F4 = Z * Z + T + N$$

Z, n, T and ' are variables While N is constant or fixed number.

Example A:

If we have a variable number 10 that varies 2 times combined with another variable number 5 with an addition or subtraction of a constant or fixed number 20; what are the possible formula representations of this problem and answer to this problem?

Solving:

$F = Z_{(1)} \times Z_{(2)} + T + N$

$Z = 10, T = 5, N = 20$

$= 10 \times 10 + 5 + 20$

$= 125$

Or

$F = Z * Z + T + N$

$= 10 * 10 + 5 + 20$

$= 125$

Or

$F = Z' + T + N$

Where $' = Z$

$= 10' + 5 + 20$

125

Or $F = Z.Z + T + N$

$= 10.10 + 5 + 20$

$= 125$

ALGEBRAIC EXPRESSION:

An algebraic expression is a combination of ordinary numbers and letters which represent numbers.

Examples:

2) $3(a**2)(b**7)$

3) (3x*y + 4*z) / (2*a - c**2)

Monomial Equation

A monomial equation consists of products and quotients of ordinary numbers and letters which represent numbers. It consists of one term.

A term is integral and rational.

Examples are:

6(x**3)y

5x / (3x**2)

-4x**3

Binomial Equation

A binomial equation is an algebraic expression that consists of two terms.

Example:

6x**2 + 7 * x * y

Trinomial Equation

A trinomial equation is an algebraic expression consisting of three terms.

Representations:

1) $F = Z_{(1)} * Z_{(2)} * Z_{(3)} + T - N$

2) F = Z**3 + T - N

3) F = Z.Z.Z + T - N

MATHEMATICAL FUNDAMENTALS

Examples:

$3x^{**}3 + 6x^*y - 2$

$3x.x.x + 6x.y - 2$

If x=2 and y=10

Then in

$3x^{**}3 + 6x^*y - 2 = 3(2)^{**}3 + 6(2)^* 10 - 2$

$\quad = 3((2.2.2)) + 6(2.10) - 2$

$\quad = (3 . 8) + (6 . 20) - 2$

$\quad = 24 + 120 - 2$

$\quad = 144 - 2$

$\quad = 142$

COMPUTATIONS WITH POLYNOMIALS

Addition:

The sum is obtained by combining rational numbers in front of the like terms.

For example: $3X^{**}2 * 2Y^{**}3 - 4x^{**}3 * y^{**}5$

$X = 2 + 3$ while $y = -3 + 5$

Subtraction:

The subtraction is done by deduction of rational numbers in front of the like terms:

Example: $3X^{**}2 * 2Y^{**}3 - 4x^{**}3 * y^{**}5$

Would be

MATHEMATICAL FUNDAMENTALS

X = 2 - 3 while y = -3 + 5

Chapter Three

APPROACHES

The approaches of solving problems or predicaments are what mathematics functions convey to the world of science. It is imperative that hypothesis be analyzed, concluded and integrated. The problem of general science is that sudden approach is realized with substances with sudden beliefs.

<u>Eliminate Method</u>

$4x + 2y = 10$ i

$0 \; - 8y = -6$...............iii

$y = -6 = -6/-8 = -3/4$

$y = -.75$

MATHEMATICAL FUNDAMENTALS

Substitution Method from equation i:

In order to find the value for x

$4x + 2y = 10$..............i

$4x + 2(.75) = 10$

$4\ xs + 1.5 = 10$

$4x = 10 - 1.5$

$x = 8.5 / 4$

$= 2.125$

Integration Method

F1: $4x + 2y = 10$..............i

F2: $2x + 5y = 8$ ii

Solving:

F1: = 4(2.125) + 2(.75)

= 8.5 + 1.5

= 10

F2: = 2(2.125) + 5(.75)

= 4.25 + 3.75

= 8

MATHEMATICAL FUNDAMENTALS 39

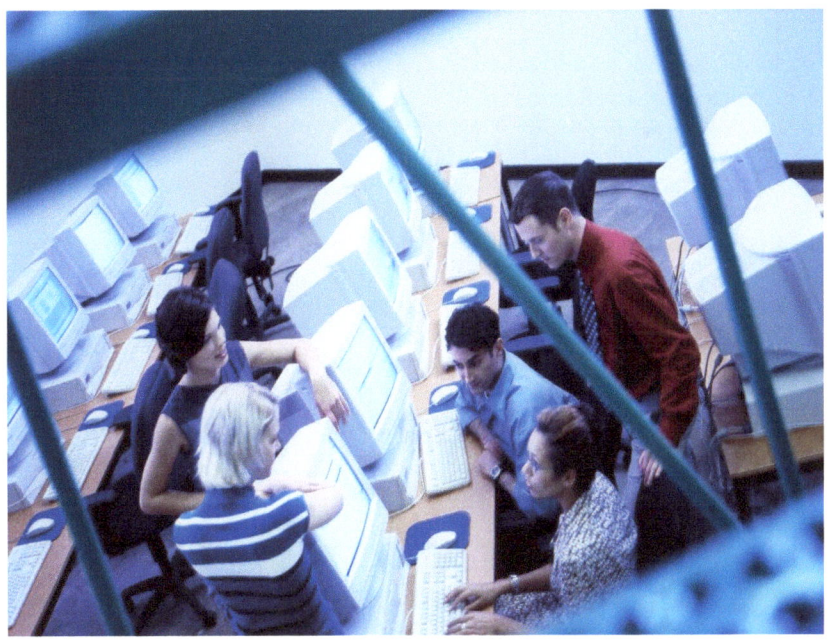

Chapter Four

APPLICATIONS

The applications of Algorithms are the willingness to control the situations around the globe. In order to sharply increase example-nary proprietaries using Algorithms we must know when, where and how to applied formula. The innovative process of mathematical Algorithms would have an impact on the behavioral of the populace. One of the technicalities of eradicating the impact of innovation is to limit and not to invigorate exposure of the formula. The applications of mathematical Algorithms may have to involve acquiring necessary skills, experience and practice. The provisions of simple and positive formula will certainly improvise the business and scientific applications as well as technical applications of Mathematics   With Algorithms we can easily solve problems precisely. Algorithms are often applied to the derivations of solutions which are not made aware to everyone.

Business Applications

We all used certain formula to carry out activities or to solve problems. Although many businesses of today use computers to solve business problems, we still have to formulate how to interpret the business solutions from these computers. Managers and business computer applications employees do formulate mathematical algorithms that are used to resolve problems manually. In the banking industries for example, mathematical algorithms are used to calculate or process customers' transactions, and in the insurance companies Algorithms are used for projections and annuity calculations.

Many large businesses depend on the use of computers to survive in the world today. The benefits of using computers to solve problems outweigh their risks. To solve problem, we must first of formulate the problem by creating algorithms or mathematical formula and then use our intuitions to manipulate, solve or process the data or the information relating to the problem. Today instead of using our intuitions, most of the businesses use computers to solve problems in all their endeavors: Accounting, Production Management, Finance, Human Resources and Operations Management and so on are some of the areas computers are applied by businesses.

Scientific Applications

Mathematical Algorithms are used by scientists for many scientific functions. They are used for predictions and solving problems of physical and unphysical entities or scenarios.

They can be used for example to derive the amount of intake of oxygen in species in a period of time. Scientists applied Algorithms in the same manner as a business man and technician or engineer. The creations of scientific Algorithms have made life easier to mankind. The use of Algorithms to solving scientist problems has made the world a better place to explore various ways of doing things and surviving on earth. Scientific applications of Algorithms actually lead to progression of human existence, and the physical being itself.

For example, formula can be used to measure the amount of intake granules a plant would need to grow.

Technical Applications

Mathematical Algorithms serve various purposes in the technical environment. Engineers just as scientists and business persons use Algorithms to carry out or perform their functions. Technical applications of formula enable engineers to design physical entities and implement most functions

The use of Algorithms is very common in engineering. For example in the designs of bridges, automobiles, robots, houses and so on, just to mention a few. An example of the usage of formula by engineer in building a bridge is to find the capacity and the length of the proposed bridge by measurement. In manufacturing automobile for example: the engineers may use formula to predict how fast an automobile can travel and the amount of gas it will consume or use.

MATHEMATICAL FUNDAMENTALS

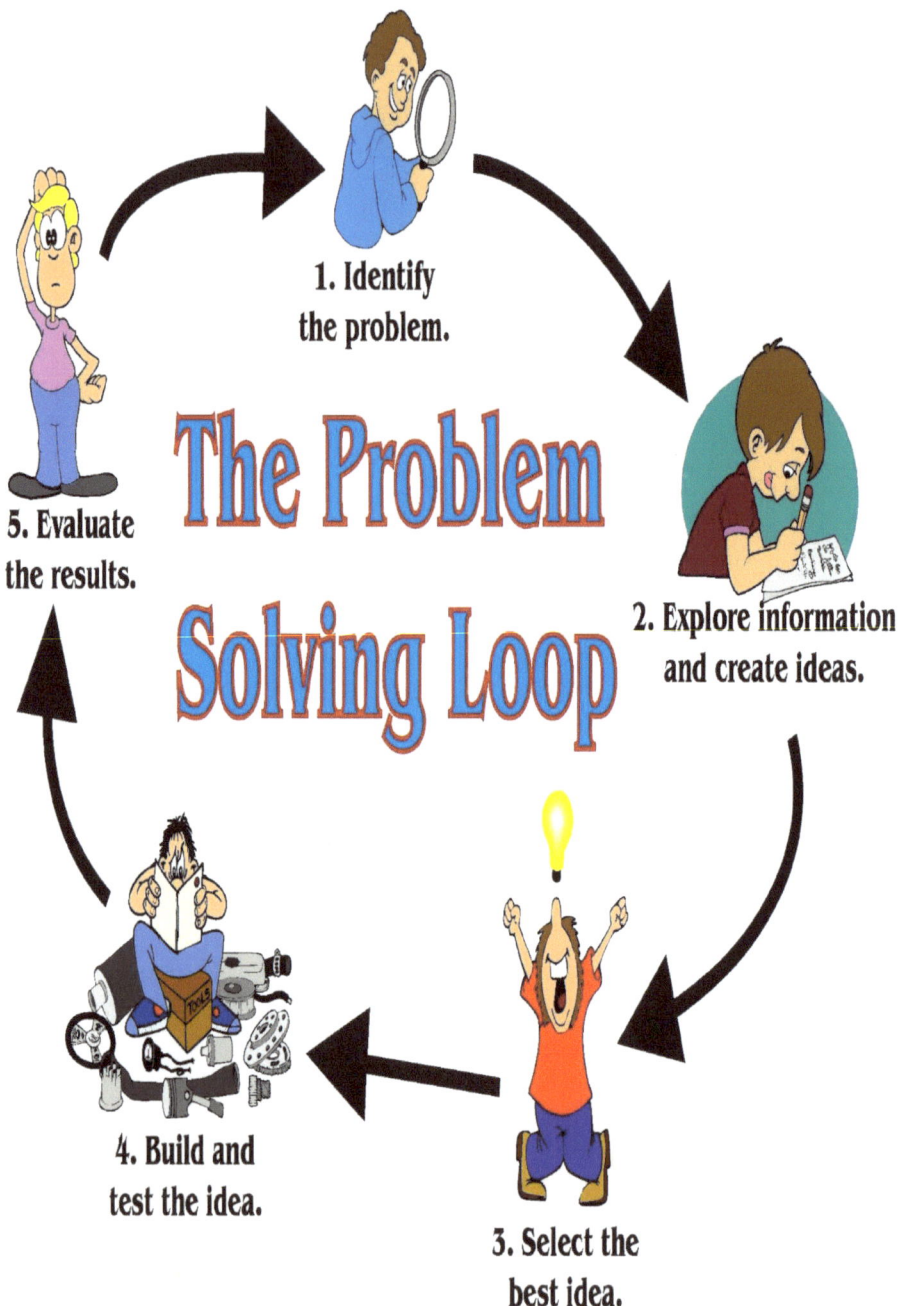

SUMMARY

PROBLEM SOLVING AS APROACHED TO EXPERT
SYSTEMS AND ARTIFICIAL INTELLIGENCE

Problem solving can be attributed to intelligent systems as well to expert systems. The systems generally play a role in behaviorism and in artifacts. The perpetuation of objects may lead to its resilience and its functionalities. The capacity to which a subject is represented depends solely on the functionalities of the issues. It has come to a place where artifacts are considered as live embodiments. Living objects and non living objects are subjected to rigorous understanding of their environment. For example, a street light knows when to turn on and turn off when it is dark or when there is a bright environment respectively. An object is subjected to scrutiny when the justifications are beyond control of behold. A form of participative objectivity is derived from subjective activities of the living objects.

The knowledge instilled in the livings can be represented as artifacts. So also can non living entities be represented as living entities, thus our world existed in learning and conveying knowledge with the aid of the living and non-living, this justification will make us believed that artificial intelligence and experts systems are both interchangeable. The complexity of acquiring knowledge is based on the sophistication of learning endeavor.

The situations of understanding lively artifacts and the process of creating physical paradigms remain a soul entity in the ages of expert systems and artificial intelligence. Approach of these subjects is subjected to the soul of soul of the era of Plato and Socrates. The breath of the mind is the elongation of physiological, psychological, sociological parabolic entities of our times. Those entities with positives and negatives thoughts can be transferred to non-livings and livings. But these means there are sensitivities between non living and the living.

This may lead us to an example of moveable objects in the world of today. Moving objects depend solemnly on the view and situation of the beyond. Transportation model, one example of serving the purpose of experts systems and artificial intelligence is dated back to the early era or years of vulnerability and adaptability which is religiously expressed in various religion peripherals, memorabilia and paraphilania.

An instructional modality may serve to accomplish a goal, and eliminate instances of confrontation between two subjects. A transportation problem analysis as oppose to agricultural and household analysis may result in the generality of instances where physical entities need to be moved between a point to another point without any disruptions, discrepancies or problems.

be moved between a point to another point without any disruptions,

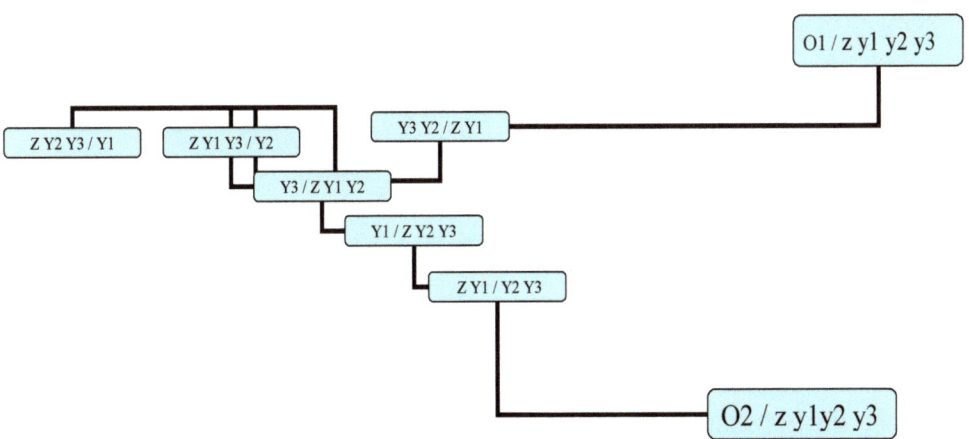

discrepancies or problems. The figure below depicts hierarchical mode of representations:

Mathematical Formula

Transportation Systems:

O_1 = Departure
O_2 = DESTINATION
/ = Boundary between the two points
Z = Sailor or Driver or Pilot or Farmer or Locksmith (INDIVIDUAL)
Y_1 = Object 1
Y_2 = Object 2
Y_3 = Object 3
(Object 1 is deadly to object 2, object 3)

BEGINNING:
$O_2 = 0$ WHEN $O_1 = Z + Y_1 + Y_2 + Y_3$

$O_1 = Y_3 + Y_2$ WHEN $O_2 = Z + Y_1$
$O_2 = Y_1$ WHEN $O_1 = Z + Y_2 + Y_3$

$O_1 = Y_3$ WHEN $O_2 = Z + Y_1 + Y_2$, $O_1 = Z + Y_3 + Y_1$ WHEN $O_2 = Y_2$
$O_2 = Z + Y_2 + Y_3$ WHEN $O_1 = Y_1$, $O_2 = Y_3 + Y_2$ WHEN $O_1 = Z + Y_1$

ENDING:
$O_1 = 0$ when $O_2 = Z + Y_1 + Y_2 + Y_3$

There are 16 possibilities of situations and 8 movements of objects and 6 occurrences of presences of the individual (Z) in this scenario.

The surveillance of destructive elements needs to be addressed to obtain a goal or to accomplish a mission.

An instance when a fisherman has a boat, a rabbit, a dog and a cat and need to transport them across a river without any fatal incident is an abominant analysis of the intelligent systems. In this scenario the fisherman cannot carry more than one thing in the boat at a time. And if the rabbit is to be left alone with the cat, there will be an altercation between the two that may lead to the dead of the rabbit. This scenario needs to be examined to determine the ways to solve a problem. The mathematical analysis of the transportation problem described previously in this text is an example of the solution to the fisherman problem.

The solutions to the fisherman problem can be discussed in a layman approach. First the fisherman needs to carry the rabbit across the river and leave the cat and the dog alone. Next he needs to leave the rabbit alone at his destination and comeback to carry the cat. He then needs to leave the cat at his destination and carry back along the rabbit.
Next thing for him to do is to carry the dog across the river and leave the rabbit alone at the source location. After he gets to his destination, he is then to leave the dog with cat and comeback to carry the rabbit. This will complete his mission without any problem between or with the species. The problem of the fisherman can also be applied to farmers; where a farmer possesses a grain, bird and a rabbit. So can the problem be applied to a locksmith with a house, a dog, solar, and mental or iron? The solar may be detrimental harmful to the dog and the mental or iron. If the locksmith is to leave the house for long period of time or to try to relocate to far away town

or country and must take with him or her no more than one item and he or she cannot leave the dog with someone or with a veterinarian and neither can he or she have somebody to keep the house in the house; then he or she will have to follow the same process as the fisherman and the farmer.

For example, if it happens that there is a Toledo or the weather gets too hot during his or her absence, the solar system may cause the mental to be too hot and thereby causing the temperature in the house to be too hot which may cause the dog to be uneasy and result in sickness or in death.

Both the locksmith and farmer problem can be resolved by following the same solution just like that of a fisherman. For the farmer with a goose, grain and cat; if the farmer leaves the goose with the grain, the goose will eat the grain and if he leaves the cat with the goose there will be problems or misunderstandings between the two which may result in the death of the goose.

To solve this problem the famer will have to take with him the goose, leave the goose at his destination and leave the cat and the grain at the initial departure, then comeback and carry the grain, leave the grain at the his destination and carry back the goose to his point of initial departure and leave goose there. He will then have to carry the cat and leave the cat in his destination with the grain and return to carry the goose.

NOW TRY TO:

USE THE ABOVE SIMILAR SCENARIOS TO SOLVE THE LOCKSMITH PROBLEM???

?

?

COMPUTERS AND ARTIFICIAL INTELLIGENCE

Artificial intelligence is the representation of human knowledge by machines or objects. Objects are devised through expert system to perform various human functions or sophisticated tasks that cannot be performed or very difficult to be performed by humans.

Computers may be referred to as artificial intelligence or expert systems.

The various usages of **Artificial Intelligence and Computers** can be enumerated as follows:

    a) Business

    b) Engineering

    c) Manufacturing

    d) Farming

    e) Mining

    f) Schools

    g) Hospitals

    h) Households.

1) **The Representation of commonsense Knowledge.** (ROBOTS)

   a) **Automobile manufacturing** (Such as in assembling and driving).

   b) **Operation services** (Such as in household/office chores, Errands deliveries.)

2) **Language Understanding.** (ROBOTS AND COMPUTERS)

   a) **Interpretation of simple questions and commands** Electronics (such as transistor radio, televisions) Manufacturing, such as in recording machines and voice recognition analyzer.

   b) Operations Services (such as in Hospitals, households and schools)

3) **Image Understanding.** (ROBOTS AND COMPUTERS)

   a) From Images to Objects Models (Such as in schools, engineering, farming, mining, hospitals and business).

   b) Computing Edge distance recognition (Such as in engineering.

   c) Interpretations of Images and surface Direction. (Such as in farming, engineering, Hospitals and business).

ARTIFICIAL INTELLIGENCE IS CONCEIVED THROUGH THE FOLLOWING MODES OF LEARNING:

1). Learning class descriptions from Samples.

2). Learning rules from Experience.

3). Learning form from functional Definition

MATHEMATICAL MODULES ARE ALSO USE TO FORMULATE THE FUNCTIONS OF COMPUTERS AND ARTIFICIAL INTELLIGENTS.

EDUCATION AND SCIENTIFIC METHODS

Education is through motivation and satisfying the needs of humans. The scientific world is part of an elongated human development. This this can be substantiated with the use and evolution of technologies. Education of the entities that comprise the need to achieve the ultimate goal of science is an important issue of today. Education is a conglomerate of beliefs of individual mind.

The scientific world plays a role in the development of education. It can be said that science is based on strategic planning. In the early evolutionary stage it has been noted through history that ideas and inventions can be obtained through exploration and scientific abomination. Education is the foundation of the continuity sustainability and transformation. The group of individual learners can be the sole of success of education of mankind.

We can achieve our needs through critical innovation of the mind regardless of our roles in society. Everyone is a learner since we do not have control over what is to be learned. The circumstances surrounding education and its mode of delivery may be due to affordability and security. These in turns affect the volatility and the flexibility of learning.

To eliminate doubts and worry, education needs to justify the prosperity of societal factors of individuals. The incumbents involve must have the resources of attaining their goals. Since we have various goals and needs, the society or organizations must always embed or include scenarios and standard of accomplishments with their expectations.

The modalities of learning comprised all entities of understanding processes of humans. The dexterity of the mind can be explained through all means of communications. Both internal and external modes of communication can be justified in the development of intelligence.

The learning processes not only consist of spiritual processes, but all physical environmental and technological scientific means. The learning modes changes as one progresses through the channel of dwelling of living. The society must realize that learning yield success only if it is applied substantially through the minds of the individual beings. Individuals learned mostly under the assumptions that they possess already all the preliminary process of life within the society. As mentioned in one of the seminar sessions in Tucson, Arizona, we as humans tend to follow with a can-do attitude. Acquiring self-knowledge always demand self-reflection. There is no way we can get to know ourselves if we don't take some quiet time to meditate. Contemplation is another one of the ways people tends to learn. Most people are willing to open to ideas and will try untested approaches and accept risk of learning. When people are at their personal best, their projects involve creative thinking and beyond-the-boundaries thinking because of the atmospheric conditions accord to them during the process of learning.

Even though we have gone through a process of learning at an early stage, we must realize that nothing is done perfectly the very first time, not in schools, not in sports, not in games and certainly not in communities. We must also understand that we as humans evolve through changes, humans tend to search for learning opportunities. Opportunities that will meet the current changes and the foresee changes. The future of science depends on the learning materials of the present.

Changes may involve physical, psychological and social changes as opposed to environmental changes in our society and schools. Education and technology rest on the hands of the beholder. Education and science must be intrigue in our mind as important aspects of life, as we progress through life. Education and science are based on needs and consequences derived from the pasts. We all make mistakes and we must learn from our mistakes which is a form of making progress.

Education and Science is based on the homogeneity of physical resources available to us as human. Our adaptation is the objectivity of our consciousness. It should be noted that contemptuous circumstances can be resolved through education.

Educating the mind is prolific; we should engage in the learning process in all avenues. Education is a process whereby we should all learn together regardless of whom you are or where you come from.

Community, identity, stability is the main characteristic of the education methodology in the society. As it is often conveyed in parabolic ways, stability is required of any individual if he or she is to succeed in the society. And for individual to portray a positive identity within the society educational stability must play a vital role in acquiring knowledge. As it is often said, Knowledge comes from learning and experience while learning and experience are respectively derived from trying and doing. Without stability and knowledge, it may be impossible to acquire success. Individual may quest for knowledge, stability and success at early stage of their educational career, but these entities may later be suppressed at a later stage of their life.

The possibility of attaining all the individual's goals may rest solely on the individual and the society at large. The learning process depends on the motivational level of the individual which may encompass the ingredients of success. The ingredient of success in the society may determine the notions of knowledge and experience. Education as an art to prowess is the basis of for integrity in the society as whole. The power of success is achievable through knowledge. The initial educational attributes of individual suffice as learning takes place. Knowledge based on experience at an initial stage may result in learning activity of the present.

Learning enterprise is a mode by which individual survived in the society. Accumulation of wealth may have to depend on knowledge and experience in the society; learning and success inevitably juxtaposed the amenities of wealth. The essence of educational training is the preparedness of individual to stability and success. It must be addressed to the problematic situations of individual in the society.

The circumstances surrounding propagation of learning is not solely materialism, but on the gratitude of knowledge. The standard which knowledge and materialism is attained is the repertoire of the educational establishments. In rationalizing the commonwealth of training individual, the society should apply transformation and sustainability in the evolution of education and science. The extenuation of objectives may depend on current and past activities. The educational solitudes may result in self-actualization of goals and thereby create self-awareness.

The technicality of learning may be justified by the scope of activities in the society. Education of the literate is different from that of illiterates in the society. Literacy does not mean everything is known, there are lessons to be learn from everyday activities in the society. Illiteracy of the mind is tolerable in certain aspect of learning.

The integration of learning may depend on the theme that individual needs, to know the anthology of survival. Stability projects the purpose of learning new ideas in our world. The determination of success rests on stability and knowledge.

Education of the mind is congenial to the cognitive approach of learning in the environment. It is believe that constant attention to the mind may gear up the learning processes. Educating the mind is a process whereby all activities are concentrated on the purposes of achieving positive results. Everyone must yield to the proliferation of the audacity to learn new ideas to attain success in the society.

There is a correlation between what is technology of the past and the present. The technological planning is based on what constitute technology in the modern society Today Information age products are based on knowledge, values and pragmatic approach. The educational approach is what constitutes technological management tools. Products management is of great importance as well as information management in globalization. The past and present terms are reciprocity of the global markets.

The retention of products serves the purpose of conservatism and pragmatic values and approach. Change of products and information are preserved in our institutions. This change is pragmatic to the management of the institutions or societies. The information age is a co-existence of values and desire. The philosophical entities composed of knowledge resources. The composition of information is the entity of products in the global market. The consistency of products is based on the magnitudes of the information obtained from the past and present.

Models and consistencies are aligned with extensive extenuation of objectivity in technology. With the extenuation of objectives, there are subjectivities to innovations. The past may be subjective to the development of the modern. Products and information are inter-changeable commodities in technological society. The aspect of globalization is empowered through local development of external entities. The adaptability of external local entity serves as philosophical global entity

The configuration of external foreign entities, such as tools of the past era results in commonwealth of technology. Technological posterity serves as the philosophical view of the modern. As a result of technological innovation there is a prowess in the dark of perspective and convenience. Information are internally and externally exposed and stored technologically. The value of information is the logical dwelling of philosophical abstracts of technology.

The transcendence of technology depends on innovation of both present and past era in order to attain sophistication and adaptability. This presents the future with lasting and endurance of educational tools in our institutions. Technology is adeptly the invigoration of educational expertise in our society. The combinations of know how, when and why will be attributed to the development and enhancement of technology tools and their awareness.

We must adhere to the improvements of the previous and the modern, as well as gear towards new developments in our institution. Education serves as a purpose for improvements of knowledge and the physical perspectives of the well-being of individual. The institutions of higher learning undergo changes in the light of technological innovations and the call for in depth knowledge of the circumstances. Thus education is a continuous and infinitely eloquent subject in our institution and in our society as a whole.

MATHEMATICAL FUNDAMENTALS

RELATIONSHIP BLOCK REPRESENTATATION

A

A1

MATHEMATICAL FUNDAMENTALS   69

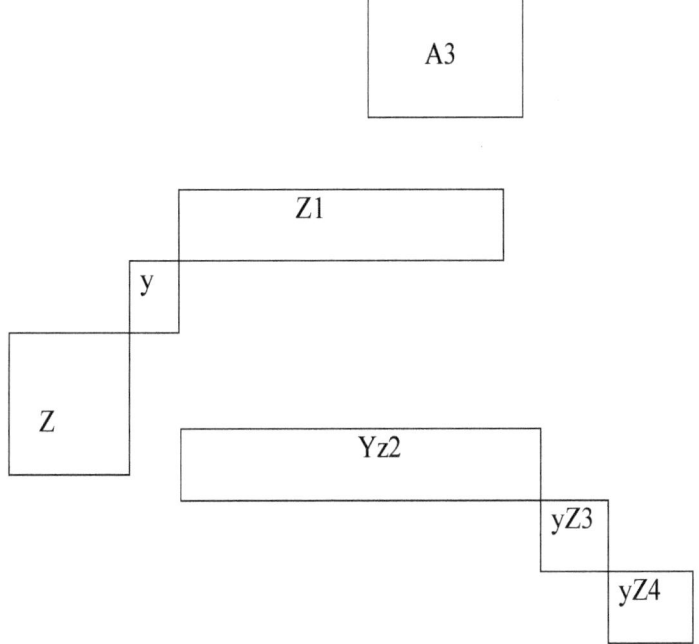

$A = (a_{1 + y1 + y3 + z4})$

$Z = (a1 + y1 + y3 + z1 + z2 + z4)$

$Y = A + Z$

13

REPRESENTATION OF DATA

Elements(x)	1	2	3	4	Total
	5	7	8	16	36 (Z)
View (y)	2	6	8	16	32 ( )
	7	8	8	16	36 (z

A					
B		C	D	E	F
Total(t)					
32		3	1	0	0

MATHEMATICAL ANALYSIS

The above chart indicates that there are four basic elements which consist of 36 different objects of which 32 are very good, 3 good, 1 fair, 0 excluded.

$$x_1 + x_2 + x_3 + x_4 = Z$$

$$y_1 + y_2 + y_3 + y_4 = z$$

$$x_1 - y_2 = C$$

$$x_2 - y_2 = D$$

$$x_3 - y_3 = E$$

$$x_4 - y_4 = F$$

$$A + B = t$$
$$Y_0 = Y_1 + Y_2 + Y_3 + Y_4 = t$$

ROTATIONAL DELAY

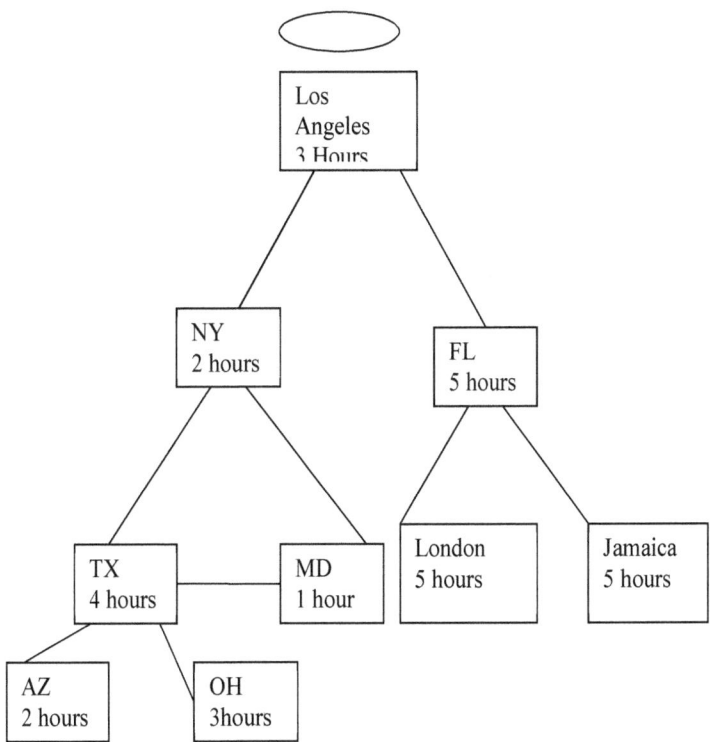

THE STRUCTURAL VIEW

Observations and Inferences in words associated with Spaceship or Plane:

Luggage	Host
Pilot	Compass
Crew	Clients/Astronauts
Space	Cloud
Sky	Air

INFERENCES AND OBSERVATIONS

Please observe and make inferences in words associated with Spaceship or Plane:

Luggage ---------Beverages----------------------

Pilot --------Helmets-----------------------

Crew ----------Flashlight----------------------

Space -----------Moon----------------------

Sky ---------Sun-----------------------

Stars ------------Dark----------------------

Instructional Development vs. Instructional Delivery

There is actually a difference between developing instructional materials for learners and delivering instructions for learners. The development of instructions may be well focused, yet the method of delivery is out focused or not well established as its form of presentations. As one develops the instruction, it is very important to stay focused on solving the performance problem when it comes to delivering the instructions. We must make sure that the objectives support the resolution of the instructional needs. One should always take into considerations either the content or skills that will help the learner improve and the mode of delivering the instructions should also be emphasized in the instructional development.

The Instructional Lecturing and Discussion Strategies

One of the most effective ways of delivering instructions is the combination of lecture and discussion method. The lecturer should always remember that learning is enhanced when learners are actively involved. In other word, participation is the key to learning process. This makes very important to get the learners involved during lectures. As the saying goes, "Get involved in it to get something from it." This means that it is therefore very important to develop a plan for including learner participation activities when lecturing. In facilitating learners' understanding of the material, lectures and discussion should also be clear and well organized.

Lectures as viewed by the majority is physically delivering of instructions or speaking to someone or to a group of individuals in a designated place, be it in a school, church or cooperate establishment. The set of instructions or statements should not be presented solely by actually speaking to audience; they should always involve both discussions and visual illustrations of topics to be lectured.

The initial presentation for an objective may provide the learner with information needed to achieve the main objective. For example, the initial presentation of a concept will include the concept name, definition and best examples – the lecturer task analysis therefore should in fact include this information.

The traditional mode of instructional delivery is by lecture. Although various modes such as distance education, in which video monitors, TV cameras or microphones, along with other optional equipment (e.g. a fax machine) is being used as the two-way transmission of voice have since evolved from traditional method of lecturing. As discussed earlier in the text, lecturing and discussion methods are instructional methods or strategies that can be used most effectively in the delivery of instructions. In lecturing, the instruction actually tells, shows dramatizes, demonstrates or otherwise disseminates subject content to a group of learners. This pattern of instructional method can be utilized in a classroom, an auditorium or a variety of locations through the use radio, amplified telephone, close circuit television transmission, interactive-distance television or satellite communication (Teleconferencing).

Strengths of Lecturing and Discussion

Since discussion offers the opportunity for a good deal of learners' activity and feedback, it should, per theory, be more effective than lecture in developing concepts and problem-solving skills. The benefits of using lecturing and discussion method to accomplish certain learning objectives can be enumerated as follows:

- Lecture and discussion format is very familiar and conventionally acceptable to instructors and learners. The lecture method is, per Morrison, Ross & Kemp (2001), the most common form of instructional delivery.

- Lectures can be easily quickly designed since the instructor is familiar with the material and will make the actual presentation. The designer of lectures usually provides the instructor with a list of objectives and a topic outline. The assumptions usually in mind is that the instructor can make necessary strategic decisions.

- Lectures and Discussion place the instructor in direct control of the class and in visible authority position. For some instructors and in many teaching contexts, the factors are advantageous for achieving the objectives.

- With a lecture, large number of learners can be served at one time. The group of learners is only limited by the size of the room; therefore, lectures can be highly economical.
- Lectures and discussions can be feasible methods of communicating when the information requires frequent changes and updates or when the information is relevant for only a short time, such as the implementation of a new travel policy.
- It should be noted that a good lecture and discussion can be motivating or interesting for the learners.

The effective use of Lecturing and Discussion

Discussions can be effectively used among matured learners and with a small group size of learners whereas lecturing is the opposite. Lecturing is most appropriate when there is a large group size of learners. Lecturing in actual sense should be augmented with discussions and illustrations.

The feelings of the learners should be taken into considerations when preparing and delivering lectures. Feeling included is the key to learner motivation to learn (Wlodkowski & Ginsberg, 1995). It is very important then that learners not encounter overt hostility, ignorance, and insensitivity, as well as more subtle messages that their cultural heritage is not valued. The message received always is that there is an underlying resentment about the pre-sentence of the learners who "don't fit." Many well-meaning but ignorant statements are the cause for discomfort in lecturing. In lecturing, we must be patient with ourselves and the learners while several generations try to unlearn some deep-seated prejudices.

While lecturing, all learners need to feel welcome. They need to feel that they are being treated as individuals. They also need to feel that they can participate fully. The need to be treated fully well is another message that needs to be addressed when delivering lectures.

One thing lecture is good for is that it can provide structures to help learners read more effectively. In fact, lecture, may help learners learn to read quickly. Readability of material depends on expectations brought to the material by the reader; therefore, better or innovative lectures can build structures and expectations that help learners read material in the given subject matter area more effectively. Lectures can also have indirect values apart from their cognitive content. Many lecturers today have motivational functions. By helping learners become aware of a problem as of challenges to ideas they have previously taken for granted, the lecturer can stimulate interest in future learning in an area. Moreover, the lecturer's own attitudes and enthusiasm have an important effect on learners' motivation. Not only is the lecturer a model in terms of motivation and curiosity, the lecturer also models ways of approaching problems, portraying a scholar in action in ways that are difficult for media or methods of instruction to achieve. Finally, there are values in lecturing for professors themselves.

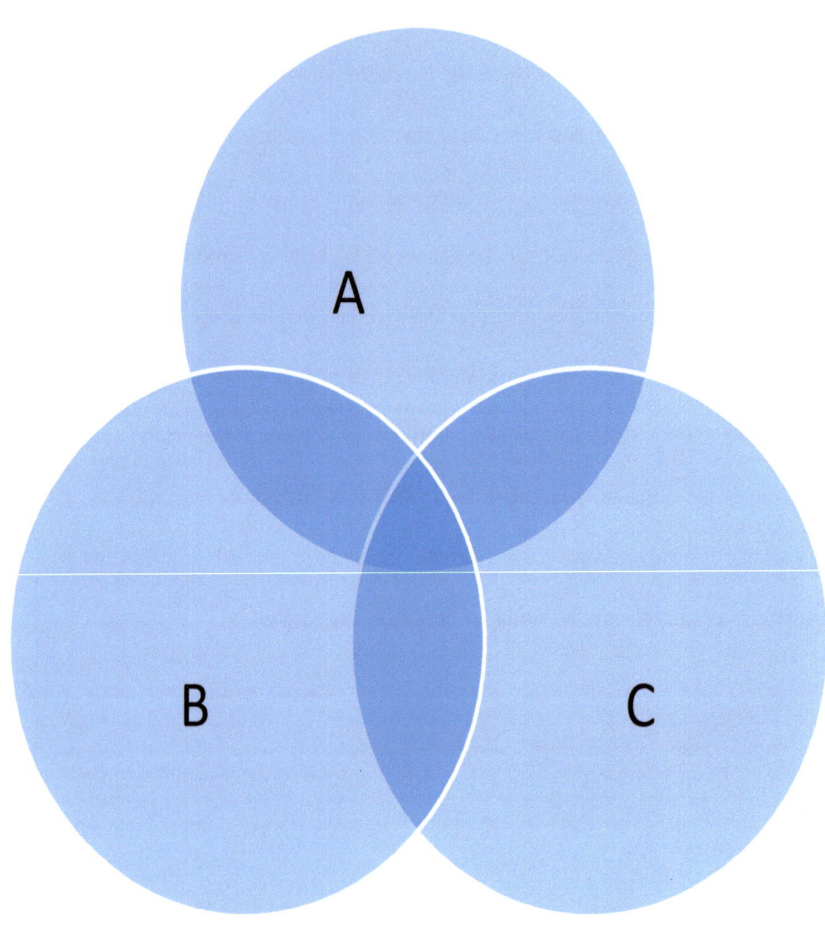

A ↔ LEARNERS

B ↔ DICUSSIONS

C ↔ LECTURES

MOTIVATIONAL CONDIMENTS OF LEARNING

Learning enterprise is a mode of which individual survived in the society. Accumulation of wealth may have to depend on the knowledge and experience in the society, learning and success extend beyond the amenities of wealth. The essence of educational training is preparedness of individual to stability and success. It must be addressed to the problematic situations of individual in the society.

The circumstances surrounding propagation of learning is not solely materialism, but on the gratitude of knowledge. The standard which knowledge and materialism is attained is repertoire of educational establishments. In rationalizing the commonwealth of training individual
have to be instituted in transformation and sustainability of the evolution of education. The extenuation of objectives may depend on current and past activities. The educational solitudes may result in self-actualization of goals and thereby create self-awareness of the episodes.

The technicality of learning may be justified by the scope of activities in the society. Education of the literates may be different from that of illiterates in the society. Literacy does not mean everything is known, there are lessons to be learn from everyday activities in the society. Illiteracy of the mind is tolerable in certain aspect of learning.

The integration of learning may depend on the theme that individual need to know the anthology of surviving. Stability projects the purpose of learning new ideas in our world. The determination of success rests on stability and knowledge.

Education of the mind is depending on the cognitive approach of learning environment and individual. It is believe that constant attention to the mind may gear up the learning process. Educating the mind is a process whereby all activities are concentrated on the purpose of achieving results. All learners may have to yield to the proliferation of learning new ideas to attain **success**.

THE KOLB LEARNING THEORY

The Kolb' theory of learning styles, which can be termed as "experiential learning" is actual based on the model of social learning. In fact, Kolb's theory came about because of the works performed by Dewey, Lewin, and Piaget. The four phases of Kolb's theory can be enumerated as follows:

(1) The learners become immensely involved in a solid experience of their own.

(2) The learners learn by reflections and observations of self and others.

(3) Abstract concepts and generalizations are symmetrically formed by the learners through inductive reasoning and not deductive reasoning.

(4) The concepts and generalizations that have been formed by the learners are empirically tested and experimented on by the learners, which eventually lead to deep involvement in solid experiences.

Per Kolb, there are four different types of individuals in the experiential learning process. These individuals can be classified as:

Divergers: These categories of individuals actually transform the experience through reflective observation by taking in the experience through solid experiences. The "divergers" view all the different aspects of a situation and then combine them into a meaningful one.

Assimilators: Abstract conceptualization is injected by these categories of individuals and it is transformed through reflective observations. The assimilators are good in creating theories and converting the diverse data into integrated whole.

Convergers: These Categories of learners take in the experience through abstract conceptualization and transform it through active experimentation. Per Kolb, these individuals are able to consume ideas and come up with the correct solutions.

Accommodators: These individual learners consume the experience through solid experience and transform it through solid active experimentation or participation. The accommodators are actually risk takers, according to Kolb, who are able to adjust to new situations and this type of learners use trial and errors methods in learning subjects or topics.

The Importance of Kolb's Contributions to Improving the Learning Process

Kolb's contributions in recognizing and providing for varying learning styles and improving the learning/teaching process are of great importance to education. Kolb's contributions actually justified the psychological and the physiological processes of learners acquiring knowledge about their ability to think and learn from experiences (Bonwell & Eison, 1991). The fours phases of Kolb's learner modes are very important to the learners' success of acquiring knowledge. Observations of self and others play a vital role in learning and teaching (Erickson, 1991). Erickson further stressed that through observations of self and others, learners are able to assimilate and converge what they learned.

Reflection is a way to convey to the individual learners what have been learnt, and it also enables the learners to make assumptions as to what is to be learnt. The Kolb's theory is a cognitive style of learning. It is also a preferred physiological modality of learning and teaching. With the Kolb's theory, the strategic learners need to be able to set

and use meaningful goals to help them generate and maintain their motivations for learning .

Providing learners with appropriate opportunities to reflect on the subjects or topics under discussion might be useful to them as they strive to reach their personal, social and occupational goals. Skills are very important in the learning process. Experiences and skills are both the conjectures of knowledge. Learners must want to learn if they are to acquire the knowledge of the subjects. According to Scott Paris and his colleagues, "Skill and Will" actually results in self-regulated learning (Paris, Lipson, & Wixson, 1983). According to an old saying, "The longevity of the learners is the homogeneity of his body, and the spontaneity of his adaptability is the objectivity of his consciousness".

KOLB LEARNING/TEACHING STYLE

Example of Kolb's Theory in an Educational Setting

An example of Kolb's theory of learning in an educational setting is where the learners engage in a laboratory work. A laboratory is set up where knowledge is put into practice. The processes of assignments or works are observed and reflected upon in a practical manner. Another example is the internship program adopted by various schools, universities and colleges. Learners are able to bring to the classrooms, experiences from the work place and to the work place, experiences from the classrooms. These methods of learning evenly provide cognitive learning styles. Testing is a way we can also relate to Kolb's learning theory. It is a way to measure what has been learnt by experience. Experience may be through studies or participating in actual activity. In an actual sense,

28

testing is an experience because the learners get to express themselves and diverge and assimilate what has been learnt.

Representation of Kolb's Learning/Teaching Theory

Fig. 1: The Learning/Teaching Modalities

The Application of Kolb Theory to Computer Instructions

The Kolb's theory of learning can be practically and adeptly applied to the computer instruction. Learners of the subjects, i.e computers, are obliged to actual experience of the practical notions of the subjects. Concepts and generalization of the topics must be made visible and adaptable so the learners can reflect, diverge, assimilate and accommodate the learned process. The learner analysis and contextual analysis should be viewed easily as identifying constraints to learning and teaching. Experience of the subjects can be attained through repetitive practice of the learned processes. Divergence and convergence of the concepts are mandatory consequences of learning the subject.